SEA AND
OF NE

This book, the third in the series, has the same specific purpose as its predecessors — namely, to help bird watchers identify the birds most commonly seen in a particular habitat. Here are illustrated birds of the shore and coastal waters. As before, species which can be confusing are grouped together, and seasonal changes in plumage are illustrated.

The text is deliberately concise, consisting mainly of features that cannot be easily illustrated and some useful facts about behaviour and relationships — species, genus, family.

Books by Elaine Power

Small Birds of the New Zealand Bush (Collins)
Waders in New Zealand (Collins)
New Zealand Water Birds (Collins)
Seabirds of New Zealand (Collins)
Elaine Power's Living Garden (Hodder)
Countryside and Garden Birds of New Zealand (Bateman)
Bush and High Country Birds of New Zealand (Bateman)

Books illustrated by Elaine Power

Tat, by Neil McNaughton (Collins)
The Horse in New Zealand, by Len McClelland (Collins)
Wild Manes in the Afternoon, by Mary Cox (Collins)
The New Guide to the Birds of New Zealand, by Falla *et al.* (Collins)
The Herb Garden Displayed, by Gillian Painter (Hodder)
A Touch of Nature, by Muriel Fisher (Collins)
Old Fashioned and Unusual Herbs, by Gillian Painter (Hodder)
Our Trees, by Frank Newhook (Bateman)
Call of the Kotuku, by Janet Redhead (Hodder)
The Pohutukawa Tree, by Ron Bacon (School Pubs.)

SEA AND SHORE
BIRDS
OF NEW ZEALAND

Elaine Power

DAVID BATEMAN
AUCKLAND

About the Artist

Elaine Power was born in Auckland in 1931, edu-
cated at Diocesan High School for Girls in Auckland
and spent a year at the Elam School of Fine Arts.
Was a librarian at the Remuera Library for five
years. After a year as a doctor's receptionist, spent
the years before marriage with the mapping depart-
ment of the Automobile Association. Married
Gerald Aldworth Power. When their third daughter
was three, started painting again and submitted
designs to the Crown Lynn Pottery Design com-
petition, in which she was highly commended. In
1968 her first book was commissioned; since then
seven of her books have been published and she has
illustrated 10 others.

© Elaine Power 1990
First published in 1990
by David Bateman Ltd
'Golden Heights', 32–34 View Road, Glenfield
Auckland, New Zealand

Typeset by Typocrafters Ltd
Printed by Colorcraft Ltd, Hong Kong

ISBN 1-86953-013-6

Contents

Introduction

Among the most remarkable and interesting of the world's birds are those of sea and shore, which range in size from the enormous 125-cm albatross to the tiny 15-cm stint. Seabirds are able to spend most of their lives at sea, only coming ashore during the breeding season; coastal birds frequent the varying habitats of the seashore.

Although New Zealand is small in area, it has an extremely long coastline with sheltered harbours, rich mudflats and sandy beaches which support a wide variety of species including gannets, terns, gulls and wading birds. Some are endemic; some travel vast distances to escape the cold winters of their northern hemisphere breeding grounds.

The waters around New Zealand extend from South Pacific latitudes to the Antarctic and provide a plentiful supply of squid, plankton, fish and crustaceans to feed albatrosses, mollymawks, petrels and shearwaters.

As island people, most New Zealanders are familiar with our coastal waters and the birds which frequent them. Some species are often observed close to land, while others are only seen by yachtsmen and deepwater sailors.

Over the years I have filled many sketchbooks with my impressions of some of these birds, from a tossing boat at sea, on the mudflats, and on windswept sandbanks. I still enjoy watching the more familiar birds, but get a real thrill when a rare visitor is found.

As with the other books in this series, this is primarily an introduction to the species most likely to be seen in a particular habitat, in this case along the shores and coastal waters of New Zealand. I hope it will provide a foundation for further and more intensive study, both of the living birds, and in the extensive literature on the subject.

I appreciate the help and advice I have received from Peter Gaze, Margaret and W. E. (Ted) Forde, and my husband Gerald, whose knowledge and skill with a word processor was invaluable when writing the text.

The birds seen in New Zealand can be categorised as follows:

Endemic	Found and breeding in New Zealand only.
Native or indigenous	Found and breeding in New Zealand, but also found in other countries.
Migratory	A two-way traffic; birds which breed here, e.g., gannets; and birds which breed in the northern hemisphere, e.g., godwits.

Books used for reference

Collins Guide to the Birds of New Zealand, R. A. Falla *et al.*, 1985, Collins.

Annotated Checklist of the Birds of New Zealand, F. C. Kinsky *et al.*, 1970, A. H. & A. W. Reed. *Amendments and additions*, 1980.

Reader's Digest Complete Book of New Zealand Birds, 1985, Reed Methuen.

The Slater Field Guide to Australian Birds, Peter Slater *et al.*, 1988, Rigby.

Because of New Zealand's length and isolation many subspecies or local races of birds have evolved. These are normally indicated by the use of trinomials, e.g., the Australasian gannet (*Sula bassana serrator*) is distinguished from the Atlantic gannet (*Sula bassana bassana*). For simplicity, however, only binomials (e.g. *Sula serrator*) have been used in this book.

PENGUINS:
Family Spheniscidae

Penguins are found only in the southern hemisphere, mostly throughout the southern oceans, though some live as far north as South Africa and the Galapagos Islands. They range in size from the emperor (*Aptenodytes forsteri*) (115 cm.) to the little blue of Australasia. All have flippers which propel them through the water and strong webbed feet for steering. There are 17 species or subspecies on the New Zealand list, many of them rare visitors from the Antarctic.

Yellow-eyed penguin — *Megadyptes antipodes*. 76 cm.

Only three species nest on the mainland of New Zealand, one of them the world's rarest penguin — the endemic yellow-eyed which breeds only on the coasts of Canterbury, Otago, Southland and Stewart Island. During most of the year their days are spent fishing at sea; at night they scramble over the rocks and beaches to roost ashore. Breeding begins in September when shallow nests are built under large logs, flax bushes and banks. The normal clutch of two white eggs is incubated by both parents, and the young, which are covered in soft brown down are fully fledged by February.

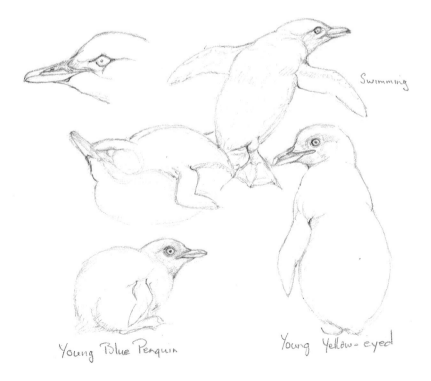

Swimming

Young Blue Penguin

Young Yellow-eyed

YELLOW-EYED PENGUIN

LITTLE BLUE PENGUIN

Little blue penguin — *Eudyptula minor.* 40 cm.

Blue penguins are common around most of the coasts of New Zealand. There are five subspecies, which vary only slightly in colour, size and shape of bills. Another subspecies is found round the coasts of south-west Australia.

They spend most of the year at sea, hunting small fish close to shore, and only occasionally coming on land, but in winter and early spring evenings pairs start to prepare for breeding, often spending the day hidden in the burrows, hollows or crevices under tree roots where they nest. At times they even nest under houses, where their noisy braying calls can be a nuisance. Two white eggs are laid and the young are covered with sooty brown down.

ALBATROSSES and MOLLYMAWKS:
Family Diomedeidae

Albatrosses and mollymawks are large, solidly built birds with long wings. They spend their lives ranging widely over the oceans, only coming to land to breed. The sexes are similar. They take all their food from the sea — fish, squid, krill — and belong, with the petrels, to the Tubinares, or tubenoses, birds which have long tubular nostrils through which they expel the excess salt taken in with their food. Of some 14 species in the world, the majority in the southern hemisphere, seven breed in the New Zealand region.

Royal albatross — *Diomedea epomophora*. 75–125 cm.

These magnificent birds wander freely over the subantarctic oceans. They often hunt fish and squid at night on the surface of the sea, but they will also follow in the wake of ships for long distances to scavenge discarded scraps.

(continued)

juvenile

ROYAL ALBATROSS

(Royal albatross continued)

The royal albatross is endemic to New Zealand, and the colony at Tairoa Head in Otago is world renowned. Adult pairs return to their nesting sites in September and remain there for the 12 months or so needed to incubate the single egg and rear the chick. The newly hatched young are covered in grey down. They fledge at 220–295 days, disperse over the Pacific Ocean and spend the next five years at sea in circumpolar travels. In the breeding season they return to the nesting sites, where they engage in social displays, but do not form pairs until mature at about nine years old.

Black-browed mollymawk — *Diomedea melanophrys*. 60 cm.

The black-browed is the most plentiful of the world's mollymawks and in the New Zealand region nests on the Antipodes and Macquarie Islands. An endemic New Zealand species, slightly smaller with minor differences of colour and plumage, breeds only on Campbell Island. Both may be seen around our coasts, where they scavenge after fishing boats to supplement their main diet of fish and krill. Like the albatross, they mate for life, and lay a single white egg.

Chick

FULMARS, PETRELS, PRIONS, SHEARWATERS:
Family Procellariidae

The members of this large family spend almost all their lives at sea. Their bodies are compact and the larger birds have long narrow wings for fast gliding flight. The smaller prions and petrels have shorter wings which enable them to flutter close to the water. All feed on fish, squid, plankton and krill, and belong with albatrosses and mollymawks to the tubenoses, which drink salt water and excrete the excess salt through their nostrils. During the breeding season the days are spent at sea, and the birds return to land at night, where most of them nest in burrows. More than 40 species have been recorded in the New Zealand region, ten of them endemic, with about 30 breeding in New Zealand waters.

Northern giant petrel — *Macronectes halli.* 65 cm.
Southern giant petrel — *Macronectes giganteus.* 65 cm.

Originally considered one species, the giant petrels have been classified separately since the 1960s. The juveniles of both have dark brown plumage, which lightens as they mature. The southern birds are quite variable,

(continued)

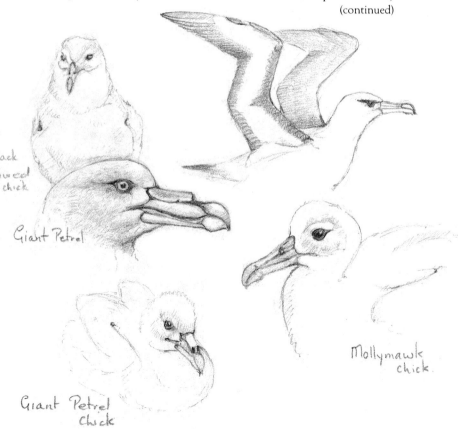

black
-owed
chick

Giant Petrel

Giant Petrel
Chick

Mollymawk
chick.

BLACK-BROWED MOLLYMAWK

(Giant petrels continued)

becoming dark with white heads or even all white; the northern is never entirely white, although the adult's head and neck lightens. Both range the southern oceans, often approaching the New Zealand coast, where they may venture into harbours to scavenge scraps from food processing factories or fishing boats. At sea they feed on fish, krill and other crustaceans and even dead birds and whales. The northern giants nest on Campbell, Antipodes, Chatham, Auckland and Stewart Islands, while the nearest colony of the southern species is on Macquarie Island.

Cape pigeon — *Daption capense.* 40 cm.

These noisy and distinctive birds are often seen in flocks offshore, where they will follow ships for scraps and even enter harbours to scavenge. They nest on cliff ledges and crevices on many subantarctic islands and lay a single white egg.

NORTHERN GIANT PETREL

PYCROFT'S PETREL

GREY-FACED PETREL

CAPE PIGEON

BROAD-BILLED PRION

BLACK PETREL

FAIRY PRION

COOK'S PETREL

GADFLY PETRELS — *Pterodroma* spp.

Medium-sized birds with wedge-shaped tails and short stout hooked bills.

PRIONS — *Pachyptila* spp.

Small petrels, blue-grey above, white below, and with dark W-shaped markings on their backs and wings. They have bristle-like lamellae on the sides of their bills which enable them to filter plankton from the water.

Grey-faced petrel — *Pterodroma macroptera.* 41 cm.
Pycroft's petrel — *Pterodroma pycrofti.* 28 cm.
Cook's petrel — *Pterodroma cooki.* 30 cm.
Broad-billed prion — *Pachyptila vittata.* 28 cm.
Fairy prion — *Pachyptila turtur.* 23 cm.
Black petrel — *Procellaria parkinsoni.* 43 cm.

Petrels and prions may often be seen near the New Zealand coast. For much of the year they cover great distances over the Tasman Sea and Pacific Ocean before returning to nest on islands off the coast of the North Island, Foveaux Strait, Fiordland and Stewart Island. They feed at night on squid, crustaceans, fish, krill and other plankton, returning to shore at night. Nests are in burrows, and a single white egg is laid. The chicks are covered in grey or brown down.

    The grey-faced petrel is also known as the North Island muttonbird.

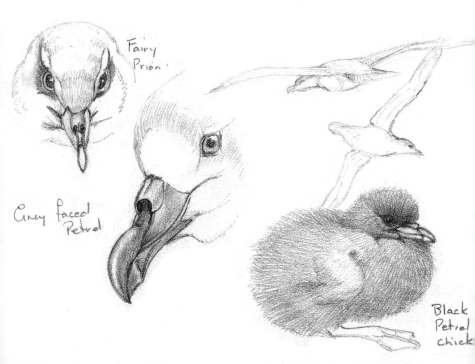

Fairy Prion.

Grey faced Petrel

Black Petrel chick

SHEARWATERS — *Puffinus* spp.

Flesh-footed shearwater — *Puffinus carneipes.* 46 cm.
Buller's shearwater — *Puffinus bulleri.* 46 cm.
Sooty shearwater — *Puffinus griseus.* 43 cm.
Fluttering shearwater — *Puffinus gavia.* 33 cm.
Hutton's shearwater — *Puffinus huttoni.* 36 cm.
Little shearwater — *Puffinus assimilis.* 30 cm.

Shearwaters are among the most superbly graceful of birds as they glide above the waves. When they dive beneath the surface in pursuit of prey their streamlined bodies and slender wings propel them rapidly through the water.

Six species are common around the coasts of New Zealand where they feed in flocks on small fish, squid and other animals which have been driven to the surface by schools of fish. Some, like the fluttering shearwater, seldom leave the New Zealand region; others, such as Buller's shearwater, migrate as far as the northern Pacific or Australia before returning to breed. All nest in burrows, and lay a single white egg. The downy chicks are grey.

Sooty shearwaters are also known as muttonbirds, and the young are still taken from nests in the Stewart Island area during April and May.

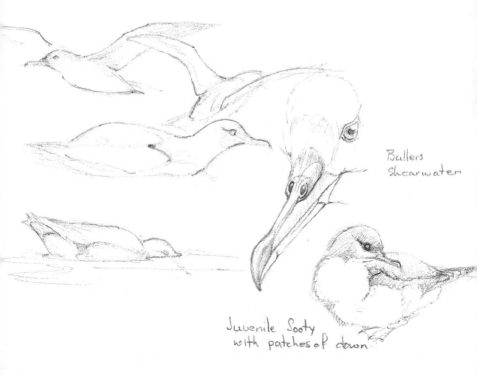

Bullers Shearwater

Juvenile Sooty with patches of down

18

FLESH-FOOTED SHEARWATER

SOOTY SHEARWATER

BULLER'S SHEARWATER

HUTTON'S SHEARWATER

LITTLE SHEARWATER

FLUTTERING SHEARWATER

STORM PETRELS:
Family Hydrobatidae

The 20 or so species of storm petrels are the smallest of the oceanic birds, ranging in size from 15 to 25 centimetres. Six have been recorded in the New Zealand region, though only two breed in the area.

DIVING PETRELS:
Family Pelecanoididae

Four species of these tubenosed swimmers are found in the southern oceans, two of them breeding in the New Zealand region.

(continued)

DIVING PETREL

WHITE-FACED STORM PETREL

DIVING PETREL

White-faced storm petrel — *Pelagodroma marina.* 20 cm.
Diving petrel — *Pelecanoides urinatrix.* 20 cm.

The white-faced storm petrel is the smallest and daintiest of the ocean birds to be seen off the coast of New Zealand. They feed mainly at night when their food of crustaceans, plankton and small fish is near the surface and hover over the water with long legs dangling as if walking on the surface. They nest on islands from the Three Kings off the coast of Northland to places as far south as the Auckland and Chatham Islands. Single white eggs are laid in burrows where the silver-grey, downy chicks are hatched. In winter they migrate to the warmth of the eastern Pacific.

Diving petrels are small chunky seabirds which skim low over the waves. They dive frequently, and their wings are as efficient for swimming under water as in the air. They will feed during the day, but more usually do so at dusk and dawn. They nest on the steep slopes of offshore islands from the Three Kings to islands off the South Island coast. The single egg is laid in a burrow which may extend underground as far as 45 centimetres. The chick is covered in silvery-grey down.

Diving Petrel. chick

Bill of Diving Petrel

very fluffy
Storm Petrel chick

GANNETS and BOOBIES:
Family Sulidae

Gannets and boobies are large streamlined marine birds. Their wings are long and pointed, the bill is strong and straight, slighly downcurved at the tip, and all have a patch of bare facial skin round the eye. Boobies are found mainly in tropical waters, gannets in temperate regions. All are expert swimmers and divers, often making spectacular plunges from a considerable height to catch their prey of small fish and squid. Of the nine species found across the world, only the Australasian gannet and the masked or blue-faced booby (*Sula dactylatra*) breed within the New Zealand region.

Australasian gannet — *Sula serrator*. 90 cm.

Gannets are often seen gliding effortlessly over the water of harbours or close to the shore searching the sea for food. They breed in densely packed colonies on many offshore islands, and a few places on the mainland, particularly on the North Island. The largest colonies are on White Island, and at Muriwai

(continued)

(Australasian gannet continued)

and the world-famous Cape Kidnappers, where they may be seen at close quarters. Most birds winter in Australian waters, returning to the New Zealand nesting sites from August onwards. Both parents incubate the single white egg; the young chick is covered with white down. They are fledged at about 16 weeks, and migrate across the Tasman with the adult birds, returning at two or three years old, though they do not breed until at least four years old.

AUSTRALASIAN GANNETS

PIED SHAG

CORMORANTS or SHAGS:
Family Phalacrocoracidae

A family with representatives in practically every part of the world. For the New Zealand area at least 14 species or subspecies have been listed, many of them endemic to a specific island group. Some are only found in coastal areas, others frequent inland freshwater lakes. Some nest in trees, others on rocky ledges. All have the habit of standing still with outstretched wings to dry themselves for they are the only web-footed birds which do not secrete waterproofing oil in their plumage.

Pied shag — *Phalacrocorax varius*. 80 cm.
Little black shag — *Phalacrocorax sulcirostris*. 60 cm.
Little shag — *Phalacrocorax melanoleucos*. 56 cm.

(see next page)

LITTLE BLACK SHAG

LITTLE SHAG

NEW ZEALAND KINGFISHERS

New Zealand
Kingfisher — *Halcyon sancta*. 24 cm.

Most likely to be seen on vantage points in coastal areas where it can find the assorted small animals that provide its food, including insects, crabs, worms, beetles, small fish and lizards. Favourite nesting sites are burrows or holes in soft cliffs and old trees.

The large pied shags are found around New Zealand and in Australian waters. They are usually to be seen in bays, harbours and coastal estuaries; only rarely on inland lakes and rivers. Nests of twigs are built in trees, very often pohutukawa, overhanging cliffs. At the start of the breeding season pied shags develop short white nuptial plumes on the crown and hind neck, and the skin on the face brightens.

The little black and the little shag are birds of both coastal and inland waters. There are two colour forms of the little shag. One is all black with a white face and neck, while the other has entirely white underparts. Little shags have a short stubby yellow bill and relatively long tail.

The plumage of the little black shag is uniformly dark, though a few white feathers appear on the sides of the head in the breeding season.

Little shags assume black crests and erect white feathers on the side of the head, and the facial skin brightens when breeding.

Both species nest in trees, and four pale blue eggs are laid in the stick nests. The chicks are naked when hatched. Juvenile little blacks are brownish-black; juvenile little shags are black, or pied — black above and white below.

Young cooling off in the heat.

Young Pied

Sketched at nesting site

Pied Shag

Young little Pied

Young little shag

chick of Pied Shag

Black shag — *Phalacrocorax carbo*. 88 cm.
King shag — *Leucocarbo carunculatus*. 76 cm.
Stewart Island shag — *Leucocarbo chalconotus*. 68 cm.
Spotted shag — *Stictocarbo punctatus*. 73 cm.

The largest of the New Zealand shags, the black shag is also known as the great or common cormorant and is found in many other parts of the world. In the breeding season the dark plumage is glossed with green and bronze, and the birds wear white plumes on neck and flanks. They are found throughout the country, in estuaries and coastal areas as well as inland waters. They roost and nest in colonies in trees, tussock and on cliff ledges, and build quite large nests of sticks, twigs and debris. Three or four greenish eggs are laid.

The king shag is an ocean bird, found in the Cook Strait area and the Marlborough Sounds. The closely related Stewart Island shag is restricted to the coastal waters of the southern South Island and Stewart Island. Both prefer to roost and nest on barren rocks, where they build large nests. At the start of the breeding season plumes appear on their heads and the facial colouring brightens. There are two colour phases of the Stewart Island shag, the bronze and the pied.

(continued)

(Shags continued)

Spotted shags are endemic to New Zealand, and are found at only a few places off the coasts of both the North and South Island. They have a very distinctive nuptial plumage with black crest and white plumes on head, neck, rump and thigh. The facial skin is strikingly bright blue and green. They nest in colonies, on rocky ledges. Three eggs are laid and both parents incubate, slowly losing their breeding dress as they do so.

SPOTTED SHAG
(juvenile)

SPOTTED
SHAG

STEWART ISLAND SHAG
(pied phase)

BLACK
SHAG

KING SHAG

HERONS and EGRETS:
Family Ardeidae

White-faced heron — *Ardea novaehollandiae.* 67 cm.
Reef heron — *Egretta sacra.* 66 cm.

Herons, egrets and bitterns belong to a large family which is represented in most parts of the world. All are birds of wetlands, both freshwater and coastal. Most herons have long legs and necks, and straight bills. They fly with slow, strong wingbeats, and heads tucked in close to the body. Their food comprises a wide variety of small fish, crustaceans, crabs, frogs, tadpoles and worms. (continued)

PIED STILTS

REEF HERON

winter plumage.
WHITE-FACED HERON

(Herons continued)

White-faced herons are immigrants from Australia. They are now very common here and frequent most coasts as well as inland pastures and lakesides, where they are often seen associating with pied stilts, which they slightly resemble, although the species are unrelated.

The reef heron is an inconspicuous, solitary bird whose grey plumage effectively camouflages it in its favoured rocky coastline habitat.

Both birds are found round the coasts of many Pacific islands, as well as Australia.

Although other members of the family nest in colonies, both the reef and white-faced herons tend to be solitary nesters. Reef herons find nesting sites in crevices, caves, under low overhanging trees, or amongst the boulders of the rocky coast. White-faced herons nest high in trees, particularly exotic pines or eucalyptus, and some of the larger native trees. Both species develop long plumes on the back in the breeding season. Three to five blue-green eggs are laid in the rather flimsy nests built of twigs. Both parents incubate; the chicks are covered in grey down.

nest buildi

Passing s to partn

Partially fledged White faced Heron chick

Reef Heron

OYSTERCATCHERS:
Family Haematopodidae

There are seven species of these neat black or black-and-white birds, of which three are found in New Zealand. The rare Chatham Island oystercatcher is found only on that group of islands. Oystercatchers feed on shellfish, gastropods, chitons, crabs, small flounder and worms. They normally pair for life and return to the same area each year to nest. Nests are mere scrapes on the ground lined with a few twigs, pebbles, or shells. Three olive-grey brown-blotched eggs are laid, and the downy chicks are buff-grey with darker stripes.

South Island pied oystercatcher — *Haematopus ostralegus*. 46 cm.
Variable oystercatcher — *Haematopus unicolor*. 48 cm.

South Island pied oystercatchers (often referred to as SIPOs) nest on the riverbeds and pastures of the South Island, but disperse to the coasts, estuaries and harbours of both North and South Islands for the winter.

   The endemic variable oystercatchers are found only around the coast, either in pairs or in the company of SIPOs. They do not migrate to other parts of the country, and do not move inland to nest. As their name suggests, variable oystercatchers are far from uniform in their colouring. Some are all black; in others the black and white pattern has a slightly blurred outline.

Chick

Oystercatcher feeding

TURNSTONES:
Family Scolopacidae

Turnstone — *Arenaria interpres.* 23 cm.

Turnstones breed in the Arctic and migrate each year to New Zealand, where, after the godwits and knots, they are the most numerous of the visiting waders. These colourful birds are usually seen in the company of other shore birds. They feed along the tideline, examining rocks, shells, seaweed and pieces of wood, deftly turning them over as they seek small molluscs, beetles and insects.

VARIABLE OYSTERCATCHERS

SOUTH ISLAND
PIED OYSTERCATCHER

summer

♂

winter

TURNSTONES

PLOVERS and DOTTERELS:
Family Charadriidae

Plovers and dotterels are members of a worldwide family, with 14 species recorded in New Zealand. Some migrate each year from the northern hemisphere, while others have become established here and now breed in some parts of the country. Two Australian species, the spur-winged plover (*Vanellus miles*) and the black-fronted dotterel have recently colonised New Zealand and are now well established. The New Zealand dotterel and the banded dotterel, the shore plover (*Thinornis novaeseelandiae*) and the wrybill are endemic.

Members of this family are boldly coloured during the breeding season, but lose their bright plumage in winter. The sexes are similar, and nests are merely scrapes in the ground where the well-camouflaged eggs and chicks merge into their surroundings. Food consists of insects, small crabs and other crustaceans, molluscs and worms.

(continued)

BLACK-FRONTED DOTTEREL

winter

BANDED DOTTERELS

summer

(Dotterels continued)

Banded dotterel — *Charadrius bicinctus*. 18 cm.

The small chunky banded dotterels are usually to be seen in little groups or large flocks. They are common throughout New Zealand in a variety of habitats ranging from lowland riverbeds, short pasture, lake edges, beaches, and mudflats to the volcanic plateau of the North Island and the mountain areas of central Otago. After the summer breeding season the birds disperse to coastal areas, and large numbers migrate to Australia, Norfolk and Lord Howe Islands. Like all plovers they perform elaborate courtship and aggressive displays, and will feign injury to draw predators away from their nests. Food consists of crustaceans, molluscs, insects and grubs.

Black-fronted dotterel — *Charadrius melanops*. 18 cm.

Black-fronted dotterels are handsome little plovers which are fairly recent arrivals from Australia. Now well established, they are found on the riverbeds of Hawke's Bay, southern regions of the North Island, and even as far south as Otago. Unlike the banded dotterels, they do not migrate. In times of danger they have a distinctive defence strategy. If an intruder approaches the nest, they will move a little distance away and pretend to feed, or will turn their backs, crouch down and merge with their surroundings. Normally three creamy eggs, heavily marked with grey or brown blotches, are laid in a shallow nest on the ground. The chicks are a tawny colour with dark markings above and white below.

Banded Dotterel

chasing intruder

Black display

threat postures

red breast displayed

Black-fronted Dotterel

Banded Dotterel chick

New Zealand dotterel — *Charadrius obscurus.* 27 cm.

The endemic New Zealand or red-breasted dotterels are found in two very different and widely separated parts of the country. In the North Island, pairs or small groups frequent shell banks, beaches and estuaries. The southern birds prefer the windswept high country of Stewart Island, from where a few disperse to the Southland coast. During the breeding season males are brown with red breasts which change to a soft buff and brown colouring over the winter months. They merge very well with their surroundings and can be quite hard to see when they are not moving about. Food is mainly crustaceans, molluscs and insects. The nests are little more than a depression in the sand dunes in the North Island, and beneath tussock on Stewart Island. Three buff or olive-green, heavily blotched eggs are laid, and the chicks are covered in buff down with black speckles.

New Zealand Dotterel.

34

Least (Pacific) golden plover — *Pluvialis fulva.* 25 cm.

Every year quite large numbers of golden plovers visit New Zealand. They arrive from Alaska and Siberia about September and spread throughout the country on to salt marshes, mudflats, around harbours and estuaries, and can be found on short pasture in paddocks. They are normally seen here in winter plumage, which is very different from their northern hemisphere summer dress. After their mottled brown and buff of winter they assume their striking breeding plumage of black and white underparts and face, with a gold tinge on the back and wings. A shy species, they do not feed in flocks, but spread out when searching for insects, larvae, beetles, caterpillars, seeds, and even crustaceans and molluscs. They have usually all left New Zealand by May.

GOLDEN PLOVER
winter
summer

NEW ZEALAND DOTTEREL

Wrybill — *Anarhynchus frontalis.* 20 cm.

Wrybills are small grey and white birds, endemic to New Zealand, and unique in being the only birds in the world whose bills turn to one side (the bird's right). They are migrants which breed only on the riverbeds of the eastern South Island and then disperse to winter on the estuaries, mudflats and harbours of the North Island where they gather in large flocks. Their dull

(continued)

WRYBILLS

CURLEW
SANDPIPER

summer
WRYBILLS

winter

RED-NECKED STINT

(Plovers and dotterels continued)

colouring merges so well with their surroundings that even large flocks can be hard to see from a distance. When disturbed they will often hop further away, often on one leg, instead of taking flight. They can be quite aggressive towards each other and will perform quite elaborate aerial displays. They are strongly territorial and will determinedly chase intruders from the nesting sites. Two pale grey eggs with blue, green and brown blotches are laid in their scrape nests. The chicks are covered in pale grey down, speckled with black.

CURLEWS, GODWITS, SANDPIPERS, TURNSTONES:
Family Scolopacidae

Curlew sandpiper — *Calidris ferruginea.* 22 cm.

Curlew sandpipers are migratory birds which breed in arctic Asia and move down to New Zealand for the winter months. They are often seen in the company of wrybills, banded dotterels and knots.

Curlew Sandpiper

Wrybills

feeding

Preening

Far eastern curlew — *Numenius madagascariensis.* 60 cm.
Asiatic whimbrel — *Numenius phaeopus.* 42 cm.
Eastern bar-tailed godwit — *Limosa lapponica.* 40 cm.
Asiatic black-tailed godwit — *Limosa limosa.* 38 cm.
Lesser knot — *Calidris canutus.* 25 cm.
Red-necked stint — *Calidris ruficollis.* 15 cm.

There are more than 80 members of the Scolopacidae family of wading birds, ranging in size from the large curlews down to the tiny stint. They all breed in the northern hemisphere, many far within the Arctic Circle, and migrate vast distances to the warmer climates of the southern hemisphere during the northern winter. More than 30 species have been recorded in New Zealand, as either regular visitors or rare stragglers. During September and October these waders start to arrive and gather in flocks on the mudflats, estuaries and shell banks from Parengarenga Harbour in the north to Invercargill in the south. In New Zealand they are usually seen in eclipse plumage of soft browns and white, though some assume breeding plumage of deep browns and reds before setting off on their return journey in April. In preparation for the flight they gather in very large, restless flocks and leave from a number of regular departure points, including Parengarenga Harbour, the Firth of Thames and Farewell Spit. A variety of marine creatures, including small crabs and other crustaceans, molluscs and worms make up their diet.

stern Curlew

Godwit

FAR EASTERN
CURLEW

ASIATIC WHIMBRE

ASIATIC
BLACK-
TAILED
GODWIT

EASTERN BAR-
TAILED GODWITS

summer

winter

LESSER
KNOTS

summer

LESSER
KNOT
(winter)

winter

x

PIED STILT

BLACK
STILT

BLACK STILT (juvenile)

STILTS and AVOCETS:
Family Recurvirostridae

Three of the world's seven species of these long-legged medium-sized wading birds are found in New Zealand. There are two stilts, one of them endemic, and one avocet — the red-necked (*Recurvirostra novaehollandiae*) — which is a very rare visitor from Australia. They all feed on molluscs, crustaceans, insects and worms.

Pied stilt — *Himantopus leucocephalus.* 38 cm.
Black stilt — *Himantopus novaezealandiae.* 38 cm.

Pied stilts are common throughout New Zealand and may be seen around most harbours, on mudflats and shell banks. During the breeding season they venture inland to nest in swampy pasture, particularly beside farm ponds, and on shingle riverbeds. The nests can vary from mere scrapes in the ground to raised platforms of grasses, sedges and roots. Four brown to olive blotched eggs are laid and the chicks are covered with buff down with dark markings. During nesting the parent birds will drive off predators, and often feign an injury or broken wing to distract intruders.

The endemic black stilt is one of the world's rarest bird species, with an estimated only 40–50 breeding in small areas of the Mackenzie Basin in the South Island. Unlike pied stilts, which are gregarious, black stilts tend to be rather solitary birds, remaining in pairs or small family groups which spend the winter months around some of the South Island lakes. The occasional bird is seen in the North Island. They take two to three years to reach maturity, during which time the plumage changes from "smudgy" to pure black. Nests, eggs and chicks are similar to those of pied stilts.

Variable black stilt

white face

hopping on one leg

Juvenile

thick fleshy legs

SKUAS:
Family Stercorariidae

Skuas belong to a family of predatory gull-like seabirds, all five species of which have been recorded in New Zealand waters, where they may be seen chasing gulls and terns to rob them of their food. The pomarine skua (*Stercorarius pomarinus* — 43 cm.) and the Arctic skua (*Stercorarius parasiticus* — 43 cm.) both breed in the northern hemisphere. They appear similar, but can be distinguished by the centre tail feathers, the Arctic being long and pointed, and the pomarine rounded. The southern great skua (*Stercorarius lonnbergi* — 63 cm.) breeds in the New Zealand region.

GULLS:
Family Laridae

Out of more than 40 gulls worldwide only three are native to New Zealand. Common in most parts of the country, they are found inland on rivers and lakes as well as on every coast. The black-backed and the red-billed are aggressive and predatory. Both sexes are similar, and nesting is usually in colonies on rocky headlands and offshore islands. Black-backed and black-billed gulls nest on dry riverbeds. Nests are built of twigs, seaweed and grasses, and the two or three eggs are usually buff-brown with dark blotches. The buff-coloured chicks have light brown markings.

Southern black-backed gull — *Larus dominicanus*. 60 cm.

Black-backed gulls are found throughout the country and as far south as the islands of the southern oceans. They have benefited from the human environment, frequenting most urban areas, and are particularly numerous as scavengers around refuse dumps. Flocks will gather on pastureland in stormy weather. Offshore they associate with terns, gannets and shearwaters when there are large schools of fish near the surface. Black-backed gulls take up to three years to reach maturity during which time their plumage changes from brown to black and white.

POMARINE SKUA
(light phase)

BLACK-BACKED GULLS
AND JUVENILES

Red-billed gull — *Larus scopulinus.* 37 cm.
Black-billed gull — *Larus bulleri.* 37 cm.

Like the larger black-backed gulls, the red-billed are to be seen on farmland and in towns and cities as well as by the coast. Offshore they fish, often in the company of terns, for anchovies and other small fish disturbed by the larger fish, but their diet includes marine invertebrates, larvae and worms as well as food scraps from picnic places and rubbish dumps.

(continued)

BLACK-BILLED GULL

RED-BILLED (juvenile)

RED-BILLED GULLS

44

(Gulls continued)

Although superficially very similar, black-billed gulls are quite different in form and habits from the red-billed. The latter mainly nest in colonies by the coasts of the sea or larger inland lakes, while the black-billed prefer to nest on the riverbeds and lake edges of inland Southland. There are also some nesting places in the North Island, particularly on the volcanic plateau. In a few places, notably Lake Rotorua, they nest with colonies of black-backed and red-billed gulls. When the red- and black-billed are seen together the more delicate structure of the black-billed is evident.

Both take from two to four years to reach maturity. Juveniles resemble the adults, apart from differences in leg and bill colouring, and greyish-brown speckling of the plumage.

Black-billed gulls are more insectivorous in their diet than the other gulls.

Tips white

White tips

soft grey.

Black-billed Gull

Red-bill Gu

Black-billed chick

TERNS and NODDIES:
Family Sternidae

Another family of seabirds found throughout the world, with 16 members recorded for the New Zealand region. Some northern terns, and the noddies (which prefer a tropical climate) are stragglers, only occasionally reaching our shores, but two species — the white-fronted and the black-fronted — are endemic to New Zealand.

Terns are elegant, slender birds with narrow, pointed wings. They swoop and glide in flight and are able to hover over the surface of the water. When fishing they will plunge from a height into the sea, seize their prey and carry it away before swallowing it.

White-fronted tern — *Sterna striata*. 42 cm.
Black-fronted tern — *Sterna albostriata*. 30 cm.
Eastern little tern — *Sterna albifrons*. 25 cm.

The most common tern of the New Zealand coast is the white-fronted, usually seen in flocks near the shore, searching for anchovies, pilchards and small crustaceans. They are also known as kahawai birds because they will follow schools of these fish and feed on the smaller fish driven to the surface during a kahawai feeding frenzy. These terns breed at three years old, making

(continued)

little Tern

dark bills
eye.

cream ——>

mottled brown
grey head.

white fronted
chick

(Terns continued)

rudimentary nests on beaches, shell banks, cliff ledges and rocky islets. Two eggs of various shades of pale blue-green with brown spots are laid, and the downy chicks range in colour from off-white to speckled brown-grey. Some immature birds migrate to Australia until they reach breeding age.

Black-fronted terns are also called inland or riverbed terns. They nest on the riverbeds to the east of the Southern Alps, and lay two stone-coloured, darkly blotched eggs in shallow twig-lined hollows in the shingle. Chicks are covered in grey down blotched with black. At this time the terns feed on insects, grubs and worms. In autumn, however, they disperse to harbours, rivermouths and estuaries throughout the country, where they feed on small fish and crustaceans.

The eastern little terns breed in South-East Asia and southern and western Australia, and are regular summer visitors to New Zealand where they may be seen foraging along the shoreline for small fish. They keep company with other shore birds at roosts around harbours, especially the Kaipara, Manukau and the Firth of Thames.

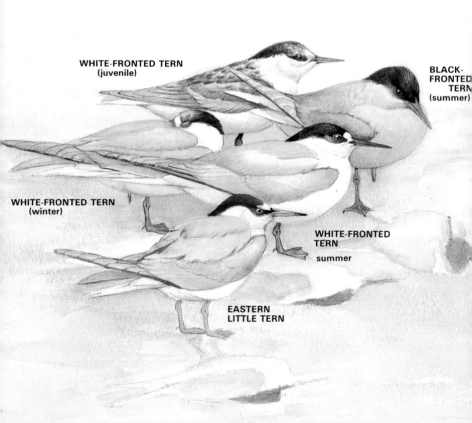

WHITE-FRONTED TERN
(juvenile)

BLACK-
FRONTED
TERN
(summer)

WHITE-FRONTED TERN
(winter)

WHITE-FRONTED
TERN

summer

EASTERN
LITTLE TERN

Caspian tern — *Hydroprogne caspia.* 50 cm.

The large Caspian terns, which are found in many parts of the world, are common throughout New Zealand, and can be found around sand and shell banks, estuaries, mudflats and harbours, often roosting in the company of gulls, oystercatchers, godwits and knots. They have adapted to living inland and are known to nest on the volcanic plateau of the North Island and on riverbeds of the South Island.

Caspian terns hunt small fish in shallow tidal waters and take small freshwater fish from lakes and rivers.

Pairs usually nest in colonies on sandy beaches and shingle banks, often in the company of other terns and gulls. The nests are just hollows in which two to three stone-coloured, dark-spotted eggs are laid. The chicks are covered with buff-coloured down with soft brown markings.

In winter the black head feathers of the adult birds fade to greyish-white.

CASPIAN TERNS

48

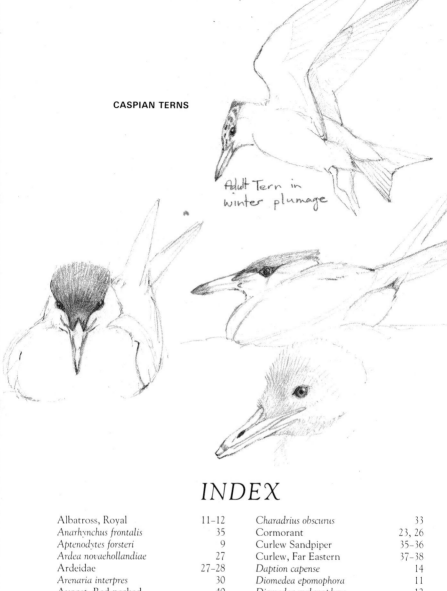

CASPIAN TERNS

Adult Tern in winter plumage

INDEX